Honey Plants of Eastern Canada

A Field Guide
for
Beekeepers, Gardeners and Naturalists

by
R. Micheal Magnini, Apiarist

Dedicated to the preservation of the natural world;

its flora,

fauna and people.

Honey Plants of Eastern Canada

A Field Guide
for
Beekeepers, Gardeners and Naturalists

Published by
Sweet Clover Productions
Scotch Lake, NS B1Y 3W4

ISBN # 0-9877299-0-x

TABLE OF CONTENTS

INTRODUCTION

For anyone interested in apiculture, insect-ecology, horticulture and/or agriculture a working knowledge of honey plants is certainly fundamentally important and practical.

While this book primarily takes the view point of the 'honey beekeeper', its principles will apply to all nectar foraging insects. All species of bees and butterflies prosper wherever there is an abundance of nectar producing plants and, a habitat, suitable for their nesting needs.

Honey plants are broadly categorized as major and minor producers of nectar. A nectar producing plant is considered a honey plant when its sugar content is large enough for insects to convert it to honey. Among these are hundreds of species across North Eastern America and scores across Eastern Canada.

Honey Plants of Eastern Canada identifies, classifies, describes and rates more than sixty species of nectar producing botanicals. The guide has flora organized as meadow plants (essential bee pasture), scrubland, marshland and woodland plants. Although woodland flora may be minor producers of nectar for honey production, they never the less contribute to the health and well-being of bee species. The variety of pollens and nectars, importantly distributed across the foraging season, strengthens the colonies for the work and labour of harvesting honey producing nectar at the peak of the summer. Inconspicuous and sometimes unknown plants can produce the finest honey.

All species of foraging insects depend upon a consistent and reliable food source. Bee pasture is one of the most abundant and reliable resources available. The action of bees ensures pollination, seed set and survival for thousands of inter-dependant species. We are one of those species.

This reference guide catalogues Eastern Canadian 'nectar producing plants' with digital photographs for clear identification, classification and description.

The Honey Plants are described by common name, scientific classification (Linnaeus), morphology (size and shape), plant type; i.e. Annual, perennial, meadow or woodland; the bloom period and the nectar/honey yield. Every honey producing plant is given a numerical rating to assign a value to the quality ~ quantity sum. That is, which plants make the best bee-pasture for the honey beekeeper.

To maintain bee health and strength, their ability to produce honey and new colonies, one must look beyond honey ....and not overlook pollen plants. These are plantae (usually trees) that produce large amounts of pollen, though without any accompanying of nectar. This is the case in many wind-pollinated plants. The pollen is usually suitable for bee consumption and often appears at critical times in the bees' life cycle, such as the very early springtime. A number of these have been included in the reference.

HONEY INDEX

0 Botanicals that produce nectar and honey of unappealing scent or taste.

1 Marginal producer of nectar/honey.

2 Low quantities of good honey.

3 Quality honey in moderate amounts.

4 Excellent honey in harvestable amounts.

5 Top class botanical. Large amounts of premium honey.

M Medicinal class. Probiotic nectar, pollen, and resin.

Honey plants of interest are divided into two classes of field crops. One being those that are considered minor honey plants due to the relatively low yields of honey acquired from them. The other is, of course, the major honey plants selected due to their great weight in honey under ideal conditions. This classification does not reflect upon taste or flavour, or other qualities that honey possesses. The Honey Rating Index divides the minor honey plants into 1) marginal production and 2) low production. Number 3 on the scale indicates plants that produce excellent quality nectar during a limited florescence. Nectar sources in category 4 render highly flavourful honeys in fairly sumptuous quantities. Class 5 references the most productive plants due to their long florescences, high output of nectar, and excellent flavours. In other words, the best plants for the honey bee keeper.

M class is used to designate plants that contain particular compounds in their nectar, pollen and/or sap that may be used in preventive medicine.

Honey production rates will vary according to weather patterns. Too much rain withers blossoms and flowers, and prevents many bees from flying as will cool or cold temperatures.

On the other hand, too little rain restricts the amount of nectar a plant can produce and thus limits the honey crops. The Honey Rating Index is based on average or normal conditions and does not represent either weather suppressed or weather enhanced environments.

The Honey Plant Guide contains the detailed descriptions listed in alphabetical order by Latin name.

A secondary directory is the Honey Plant Calendar presented in chronological order from the first flower of spring to the last one in autumn.

Some flowering plants, such as the buttercups, daisies, lupines, spirea, and garden loosestrife Lysimachia quadrifolia have not been included for the simple fact that honey bees do not feed from these flowering plants. Their pollination, if required for reproduction, is conducted by other vectors such as ants, flies, moths and/or the wind.

Many listed honey plants are not native or indigenous species of Canada, but are transplants brought to America from the Old World by European settlers. However, many of these plants are naturalized and effectively adapted in terms of honey bee (also not native to North America) ecology and honey production.

This reference contains all the major honey plants in Eastern Canada. Undoubtedly some minor honey plants have been bypassed, such as the 'partridge berry' of Newfoundland, and a few garden flowers and conventional farm crops, thus expansion of this field guide will be necessary. Meanwhile, I hope this book will assist you in managing good and healthy bee-pastures.

I would like to dedicate this guide to the bees of the world, our families and friends, who live in harmony with each other and are blessed by their company.

BEE ECOLOGY

BEES are a keystone species. As the name implies, these insects are fundamentally important to the unity of the biological system involving related and unrelated species.

That is, without their natural influence in the world, complex biological systems would suffer deterioration, and collapse into a simpler, less complex form of bio-sphere. A return to the desert. To remove the bee (or worse, all pollinating insects) would cause the number of inter-dependant species in an ecosphere to cascade into oblivion and extinction.

It is absolutely necessary to have a healthy bee population and a population of healthy bees in order to have a robust biological environment. This diverse and dynamic environment is essential to the well-being of all living creatures.

Successful beekeeping comes from a keen understanding of the ecologies of honey bee colonies, flowering plants, predators and pathogens.

For bees to succeed in their 'pollination mission', their needs in regard to nest building, food gathering, reproduction and seasonal acclimation must be met. Thousands of flowering plants depend upon external 'carriers' of their pollen to fertilize the waiting ovum of their sisters. These waiting flowers may be centimeters away or kilometres apart. Older species rely on wind to accomplish this goal, while the more recently evolved species have specialized into attracting insects (and in some cases birds) to transfer their precious genetic material. By producing nectar -- a sucrose rich liquid -- in their nectaries, they offer food to the insects in exchange for pollen transfer. In nature, it is an equitable trade. The plants' concentrated efforts results in direct and efficient pollination, and seed set which begets continued generations of the species. For the insects, they are guaranteed a sustainable food source.

While bees are not 'aware' of their function and importance as pollinators they none-the-less are most prolific and dedicated to the task. While a large number of insects make use of nectar and pollen as food, only bees (all species) make honey from plant nectar. Since honey bees store honey for later use, feeding young and winter survival they must convert the nectar to a more concentrated form. They do this by first dehydrating the nectar to concentrate the sucrose. This is followed by an enzymic process within their gut (honey-stomach) which converts the sucrose to fructose and glucose, as well as incorporating a number of anti-biotic enzymes such as glucose oxidase, invertase and lacto-bacillus. These tend to preserve the honey from bacterial and fungal spoilage.

While nectar and its derivative, honey are eaten by insects, especially bees, nectar lacks proteins. Thus the bees, and other insects, have taken advantage of the plants' pollen as a source of protein, vitamins and minerals. Bees have evolved special hairs on their bodies to which pollen grains will adhere. Honey bees have developed 'pollen baskets' from long, curved hairs on their rear legs. In comparison, wasps have smooth ectoskeletons, without hairs, since they are primarily carnivores. In the case where bees are feeding on pollen, the plants have had to respond by producing copious amounts of pollen to ensure pollination and survival. As each species adapts to new opportunities, other interdependent species must continue to change, to evolve, and flourish in new environments. The invisible force of evolution is always in action, albeit slow in our eyes.

According to Karl von Frisch, bees exhibit predictable behaviour as determined by their sensory abilities, communications and learning. Bees have essentially the same senses as humans, touch, hearing, sight, taste and scent; however they exist in different proportions and sensitivities. The primary sense that bees possess and utilize is scent. With their highly sensitive antennae they scout the flora of the countryside to find the richest sources of nectar, and are able to pass this information on to the legions of foragers in the colony. The colony itself is organized around a series of 'pheromones' and other chemical indicators produced by the queen, the larvae, the workers and drones. And thus they are easily disturbed by the odour of animals, people, and synthetic chemicals. Honey bees are well endowed with the ability to communicate by means of scent, sight, touch, taste and sound.

These abilities have been well documented through considerable scientific experimentation.

For those interested in studying the details of these remarkable phenomenon, I highly recommend the book, "*The Dancing Bees*" by Karl von Frisch.

The primary ecological action of bees is the pollination of plants producing seeds and fruits. The bees' motivation is selfish in that they seek only to feed, reproduce and live. Likewise, the production of nectar and pollen serves the plants to the same end. However, a cascade of life forms benefit from this simple primary action. Herbivores of all kinds from the moth or butterfly caterpillar to the large mammals such as deer, moose and horse are provided with abundant, nutritious resources. Likewise, the predators of these animals, from spiders to eagles, prosper in the rich environment caused by the interaction between plants and pollinating insects. It should be noted too, that mankind is a beneficiary of this environment and ecology through the production of a large portion of their agricultural produce. Many common foods, such as tomatoes, peppers, apples, beans and peas would scarcely exist were it not for the family of bees.

To contemplate life without bees would envision a desperate world suffering from widespread famine.

It should now be easy to see the importance of a bio-diverse bee-pasture. It is us who recognize the 'bigger picture', beyond the active players in this life drama, who understand the inter-connectivity of organic life and must be responsible for our actions. The power of mankind to re-shape the Earth has reached its apex, and only the sane and reasonable application of this power and knowledge will deliver a sustainable and prosperous future.

Here then, by implication, the importance and necessity of identifying nectar plants, maintaining their habitat, and even the action of proliferating bee pasture is clear. To prosper as a beekeeper, a farmer, a gardener, or a homeowner the knowledge of nectar/honey plants is vital and sacred.

HONEY PLANTS OF EASTERN CANADA

Botanical Names

ALFALFA	Medicago sativa
ALSIKE CLOVER	Trifolium hybridum
APPLE	Malus pumila
APRICOT	Prunus armeniaca
ASTER	Aster novi-belgii
ASTER	Aster umbellatus
BEAUTY BUSH	Kolkwitzia amabilis
BLACKBERRY	Rubus fruticosus
BLUEBERRY	Vaccinium angustifolium
BORAGE	Borago officinalis
BUCKWHEAT	Fagopyrum esculentum
BURDOCK	Arctium minus
CANADA THISTLE	Cirsium arvense
CHICORY	Cichorium intybus
CHIVES	Allium schoenoprasum
CHOKE CHERRY	Prunus virginiana
COLTSFOOT	Tussilago farfara
COMMON MULLIEN	Verbascum thapsus
CRANBERRY	Vaccinium macrocarpon
CROCUS	Crocus vernus
DANDELION	Taraxacum officinale
DEWBERRY	Rubus spp.
DOGWOOD	Cornus alternifolia
FIREWEED	Epilobium angustifolium
GARDEN POPPY	Papaver spp.
GLADIOLUS	Gladiolus officinalis
GOLDENROD	Solidago spp.
HANSA ROSE	Rosa rugosa 'Hansa'
HARDY SHRUB ROSE	Rosa rugosa spp.
HAWKWEED	Hieracium aurantiacum
HAWTHORN	Crataegus monogyna
HENRY HUDSON ROSE	Rosa rugosa 'Henry Hudson'
HEATH	Erica carnea
HORSE CHESTNUT	Aesculus hippocastanum
JAPANESE KNOTWEED	Fallopia japonica
JOE-PYE WEED	Eupatorium maculatum

LAVENDER	Lavendula spica
LILAC	Syringa vulgaris
LOCUST (Black or Honey)	Gleditsia spp.
MONEYWORT	Lysimachia nummularia
NORTHERN CATALPA	Catalpa speciosa
OREGANO	Origanum vulgare
PEACH	Prunus persica
PEAR	Pyrus communis
PLUM	Prunus domestica
PUMPKIN	Cucurbita maxima
PURPLE LOOSESTRIFE	Lythrum salicaria
QUEEN ANNE'S LACE	Daucus carota
RED CLOVER	Trifolium spp.
RED MAPLE	Acer rubrum
RHODODENDRON	Rhododendron spp.
ROUGH-FRUITED CINQUEFOIL	Potentilla recta
ST. JOHN'S WORT	Hypericum perforatum
SERVICEBERRY	Amelancherier laevis
SCOTCH HEATHER	Calluna vulgaris
SCOTCH THISTLE	Cirsium spp.
SHOWY MOUNTAIN ASH	Sorbus decora
SPECKLED ALDER	Alnus rugosa
STITCHWORT	Stellaia graminea
STRAWBERRY	Fragaria virginiana
SUGAR MAPLE	Acer saccharum
SUNFLOWER	Helianthus annuus
SWEET CHERRY	Prunus avium
SWEET ROCKET	Hesperis matronalis
TEA ROSE	Rosa multiflora
THYME	Thymus vulgaris
TUFTED VETCH	Vicia cracca
WILD ROSE	Rosa virginiana
WHITE CLOVER	Trifolium repens
WHITE SWEET CLOVER	Melilotus alba
YELLOW ROCKET	Barbarea vulgaris
YELLOW WOODSORREL	Oxalis stricta

Acer rubrum

Acer saccharum

RED MAPLE *Acer rubrum*

TYPE:	Woodland, perennial.
HABITAT:	Rich, moist soils along stream beds and wetlands.
DESCRIPTION:	Very large tree often divided and spreading. Bark is smooth grey to reddish brown turning dark grey with large scales. Leaves are simple, opposite 3 to 5 lobed with sharp, V-notches, turning brilliant red in autumn.
FLOWER:	April / May before the leaves, unisexual and bisexual in small red clusters. Predominately wind pollinated.
ZONE:	4 to 6
HONEY INDEX:	1 No known samples of *Acer* nectar. Red pollen visible and abundant in and around brood nest. Most valuable as a source of very early spring feeding for young bees.

SUGAR MAPLE *Acer saccharum*

TYPE:	Woodland perennial.
HABITAT:	In pure stands or mixed forest on moist, rich, well-drained soils; often along streambeds.
DESCRIPTION:	Very large deciduous tree, wide and spreading in open spaces, narrow and tall in crowded woods. Leaves are simple, opposite and five-lobbed with sharp points. Bark is smooth and grey, becoming rough, furrowed and fissured.
FLOWER:	Occur May / June with leaves; are unisexual and bi-sexual in small greenish clusters. Predominately wind pollinated. Bloom period approximately two weeks.
ZONE:	4a to 6
HONEY INDEX:	1 No known samples of nectar. Greenish pollen abundant and stored throughout the hive.

Aesculus hippocastanum

Allium schoenoprasum

HORSE CHESTNUT *Aesculus hippocastanum*

TYPE: Woodland, perennial.

HABITAT: A non-native tree usually found as an ornamental landscape specimen on lawns, parks and golf courses. Aesculus grows in a variety of soils within mixed hardwoods.

DESCRIPTION: Very large tree with large, pendulous leaves that are alternate and wide near the tip, pale green and toothed. A minor or major honey plant - depending upon size and numbers - in the northeast. It is very attractive to all types of bees.

FLOWER: Huge spikes of colourful blossoms, mostly white, appear after the leaves in June. Unisexual and bisexual flowers appear on the same tree, bloom lasting for two to three weeks.

ZONE: 5a to 7a

HONEY INDEX: 3

CHIVES *Allium schoenoprasum*

TYPE: Meadowlands, perennial.

HABITAT: Herb gardens, flower beds, naturalized landscapes. Requires full sun on slightly acidic, well-drained soils that are moderately rich and aerated.

DESCRIPTION: Leaves are tubular as well as floral stalks. Grows in dense clumps to 45 centimeters in height.

FLOWER: Clusters of tiny, six-petalled flowers, 1 cm across, atop of tubular stalk, 30 - 40 cm in height. Florets are mauve, purple and pink depending on variety. Blooms in June lasting four weeks or more.

ZONE: 4b to 7a

HONEY INDEX: 2 (M)

Alnus rugosa

Amelancherier laevis

SPECKLED ALDER *Alnus rugosa*

TYPE: Woodland, perennial.

HABITAT: Wet, well-drained sandy or gravelly soils along streams, swamps and lakes in full sun.

DESCRIPTION: A low, crooked, spreading tree shrub up to four meters in height, growing in clumps and thickets. Bark is smooth, brown with conspicuous white pocks. Leaves are alternate, simple, oval, sharply pointed and double toothed.

FLOWER: March / April unisexual in clustered reddish brown catkins opening before the leaves. Male and female flowers appear on the same twig. Bloom period two to three weeks. Predominately wind pollinated.

ZONE: 1a to 6b

HONEY INDEX: 1
A very valuable source of pollen for bees in the early spring.

SERVICEBERRY *Amelancherier laevis*

TYPE: Woodland, perennial.

HABITAT: Open areas in mixed woods and old pastures.

DESCRIPTION: Small deciduous tree has simple, alternate leaves and slender twigs with smooth, greyish-brown bark. Older bark is rough and scaly. Small fruits resembling blueberries are edible.

FLOWER: Clusters of white, bisexual blossoms appear before and with leaves. Profuse flowering covers branches lasting two to three weeks. Bloom occurs in early spring.

ZONE: 3a to 6a

HONEY INDEX: 1

Arctium minus

Aster novi-belgii

BURDOCK *Arctium minus*

TYPE:	Perennial, woodland transitional.
HABITAT:	Roads, paths and trails, fields and woods on mixed soils.
DESCRIPTION:	Heavy, tall, branched plant, 2 m, with large tobacco-like leaves covered in hairs. Deep tap root supports thick stalk 6 - 7 cm diameter. A major honey plant in forests and fields.
FLOWER:	Purple flower head of multiple florets emerging from a globular, green seed-pod covered with stiff 'hooked' hairs. Florescence occurs July and August.
ZONE:	5a - 7a
HONEY INDEX:	3

ASTER *Aster novi-belgii* *Asteraceae*

TYPE:	Perennial, meadowlands and woodland.
HABITAT:	Prefers open spaces in fields, woods and roadsides on damp ground.
DESCRIPTION:	Stiff stemmed plant with simple lancelet leaves varies in size and height, up to two meters. Clusters of plants appear bushy. A major honey plant which produces fine late season honey.
FLOWER:	Profuse florescence of blue petalled flowers, 2 to 3 cm across with yellow center. Blooms August to October.
ZONE:	4b to 6b
HONEY INDEX:	4

Aster umbellatus

Barbarea vulgaris

ASTER

Aster umbellatus
Asteraceae

TYPE:	Perennial, woodland.
HABITAT:	Marshes, roadside ditches, damp woody areas in sunny locations.
DESCRIPTION:	Stiff stemmed plant with multiple branches on upper portion. Lanceolate leaves are small with smooth edges. A major honey plant, producing late into the season.
FLOWER:	Profuse flowering of small, white petalled flowers, 2 cm diameter, with yellow centers. Florescence occurs late August to October.
ZONE:	4b to 6b
HONEY INDEX:	4

YELLOW ROCKET

Barbarea vulgaris
Cruciferae

TYPE:	Annual, meadowlands.
HABITAT:	Open fields, ditches, and pastures on medium soils.
DESCRIPTION:	Invasive member of the mustard family, 25 to 50 cm in height. Has a central stalk with alternate, narrow leaves. Not an insignificant honey plant when abundant.
FLOWER:	Clustered on spikes along length of main stalk. Bright yellow small flowers bloom through May and June. Attractive to bees.
ZONE:	4b to 7a
HONEY INDEX:	3

Borago officinalis

Calluna vulgaris

BORAGE

Borago officinalis
Boraginaceae

TYPE:	Meadowlands, annual.
HABITAT:	In full sun on rich, well-drained soils.
DESCRIPTION:	A sprawling herbaceous plant 20 to 100 cm in height with 5 to 15 cm long alternate, simple leaves. Stalk and stems are very hairy.
FLOWER:	Blue star-shaped flowers with five pointed petals and black centers. Each branch has numerous, drooping, individual flowers on stems along both sides. Florescence begins in July and continues until the frost (October).
ZONE:	5a to 7a
HONEY INDEX:	4 (M)

SCOTCH HEATHER

Calluna vulgaris

TYPE:	Perennial, scrublands.
HABITAT:	Open areas, rock gardens on acidic, peaty soils.
DESCRIPTION:	Short, spreading shrub, 25 to 30 cm, with many thin branches covered in small, needle-like leaves, 1 cm long. Identical in appearance to Heath (Erica carnea) differing only in the bloom period.
FLOWER:	Numerous, small, bell-shaped flowers, whitish-pink to red in colour covering upright stems. Blooms in late summer (August) for two to three weeks.
ZONE:	4b to 6b
HONEY INDEX:	3

Catalpa speciosa

Cichorium intybus

NORTHERN CATALPA *Catalpa speciosa*

Bignoniaceae

TYPE:	Perennial, woodland.
HABITAT:	Lawns, parks and golf courses on rich well-drained soils. Prefers full sun. A non-native transplant species.
DESCRIPTION:	Large tree, up to 40 meters in height. Broad, pale green, heart-shaped leaves and soft pith-like wood. Bark is brown and furrowed on trunk. Long, thin seed pods remain on tree over winter. A minor honey plant in Canada that produces premium quality honey when trees are abundant.
FLOWER:	Very similar in appearance to acacia (locust) trees, with white, 2 to 3 cm flowers in clusters on long stems. Blossoms appear after the leaves in July. Florescence lasts approximately three weeks.
ZONE:	5b to 7b
HONEY INDEX:	4

CHICORY *Cichorium intybus*

TYPE:	Perennial, meadowlands.
HABITAT:	Roadsides, ditches, fence lines, abandoned fields on coarse, gravelly soils.
DESCRIPTION:	Deep tap-root supports tall, tough stems up to one meter, with a rosette of deeply toothed, lanceolate leaves near the ground. A few smaller leaves ascend the stem with flowers. Widely distributed, naturalized transplant species of Europe.
FLOWER:	Large, pale-blue flowers, 4 cm across with numerous narrow petals, scattered along tall stems at leaf nodes. Late summer bloom lasts for eight to ten weeks.
ZONE:	5a to 7a
HONEY INDEX:	1

Cirsium arvense

Cirsium spp.

CANADA THISTLE

Cirsium arvense
Asteraceae

TYPE:	Perennial, meadowlands.
HABITAT:	Open fields, roadsides, rough infertile ground in full sun.
DESCRIPTION:	Consisting of numerous stalks from a central tap root plant reaches 1 m in height. Leaves are irregular with tiny sharp hairs. A major honey plant, nectar and pollen are abundant and sustained producing excellent summer honey.
FLOWER:	Single, purple flower head of multiple florets at top of stalk. Numerous flower stalks per plant. Florescence period July through September.
ZONE:	4b to 7a
HONEY INDEX:	5

SCOTCH THISTLE

Cirsium spp.
Asteraceae

TYPE:	Annual, meadowlands.
HABITAT:	Fence rows, roadsides, open fields on marginal soils.
DESCRIPTION:	Bright green and shiny plant up to one meter in height. Jagged leaves and stems are covered with sharp, hard ‘thistles’ or thin thorns. Often grows in clumps or thickets.
FLOWER:	Spiky blue or purple florescence on individual stems emanating from a central stalk. Blooms in late summer for two to three weeks. Produces thistledown to carry seeds on wind.
ZONE:	4b to 7a
HONEY INDEX:	3

Cornus alternifolia

Crataegus monogyna

DOGWOOD *Cornus alternifolia*

TYPE: Perennial, woodland.

HABITAT: Prefers cool, shady woods on moist soils.

DESCRIPTION: Small tree with a low spreading crown of horizontal branches. Many smooth upright twigs that are alternating as are the leaves which are oval with wavy edges. The bark and leaves have purplish streaks. The fruit is small and blue.

FLOWER: Small, cream-white bisexual flowers appear in showy clusters near the ends of branches. Petal-like bracts are missing from this species' flowers. Blooms briefly in June.

ZONE: 4a to 6b

HONEY INDEX: 1
Extremely attractive to wild, native bees.

HAWTHORN *Crataegus monogyna*

TYPE: Perennial, woodland scrublands.

HABITAT: Prefers open areas on rich, moist, well-drained soil.

DESCRIPTION: Small, dense tree with abundant long thorns growing in thickets. Brown bark, deeply, sharply toothed leaves are simple and shiny.

FLOWER: Clusters of small white blossoms appear after leaves in June. A sour odour is produced in the nectar which remains in the honey of this tree. Attracts bees.

ZONE: 4a to 7a

HONEY INDEX: 0
CAUTION – Produces foul smelling / tasting honey.

Crocus vernus

Cucurbita maxima

CROCUS *Crocus vernus*

TYPE:	Perennial, meadowlands.
HABITAT:	Rock gardens, flower beds and lawns.
DESCRIPTION:	Short, perennial garden flower grows from shallow bulbs in a wide range of varieties. Narrow, upright leaves appear after bloom.
FLOWER:	Large, showy, brightly coloured large-petalled blooms appear in April, just above ground level, sometimes poking through spring snow. Bloom period is two to three weeks.
ZONE:	4b to 7a
HONEY INDEX:	1 First true nectar plant in spring.

PUMPKIN *Cucurbita maxima*

TYPE:	Annual, field crop.
HABITAT:	Light, loose, fertile and rich soil in open areas with full sun. Requires good drainage and prospers in compost.
DESCRIPTION:	Ground creeping vine with large, arrow-shaped leaves on upright stalks, 30 to 60 cm in height. Hollow stalks and leaves covered in fine hairs. Sprawling vines reach 10 to 20 meters across.
FLOWER:	Very large, brilliant yellow / orange flowers, 10 to 12 cm across have four petals. Unisexual blossoms on upright stalks with male and female flowers on same plant. Large nectaries attract many bees.
ZONE:	5a to 7a
HONEY INDEX:	3

Daucus carota

Epilobium angustifolium

QUEEN ANNE'S LACE

Daucus carota
Apriaceae

TYPE:	Biennial, meadowlands.
HABITAT:	Dry open fields, roadsides on rough, well-drained ground.
DESCRIPTION:	Tall, thin plant 1.5 m, with sparse fern-like leaves. The flowering head has a dark purple center. Stalks covered in fine hairs. Very attractive to many bee and wasp species however with limited nectar. A minor honey plant despite its numbers and size.
FLOWER:	Tiny white flowers, 4 to 5 mm, cluster into a wide flat head at top of narrow, stiff stalk. Florescence period July to September.
ZONE:	4a to 7a
HONEY INDEX:	1

FIREWEED

Epilobium angustifolium
Onagraceae

TYPE:	Perennial, meadowlands, open woodland.
HABITAT:	Open, disturbed ground along roadsides, farm fields, and burnt forest land on mixed soils.
DESCRIPTION:	Tall, single stemmed plant, 1 to 2 meters height, grows in thick clusters covering large areas of new ground. Leaves are lanceolate, 15 cm long, alternate and dark green with smooth edges. A major honey plant where abundant.
FLOWER:	Tall spike of pale purple flowers, 2 cm across, at top of main stalk. Flowers appear sequentially, blooming up the main flower stem. Florescence begins in mid-July and continues until September.
ZONE:	4a to 6a
HONEY INDEX:	5 (M) A premium honey with distinctive, spicy flavour and medicinal qualities.

Erica carnea

Eupatorium maculatum

HEATH

Erica carnea
Ericaceae

TYPE: Perennial, meadowlands.

HABITAT: Rock gardens, landscapes and flower beds on peaty or acidic soils.

DESCRIPTION: Small to medium sized shrub with many thin stems covered in needle-like leaves. Hardy plant in windy, wet and cool conditions.

FLOWER: Numerous, small bell-shaped flowers crowed along thin stems. Pink or white colouration, bloom period is early spring lasting two to three weeks. Extremely attractive to bees in early springtime.

ZONE: 4a to 6a

HONEY INDEX: 3

JOE-PYE WEED

Eupatorium maculatum
Asteraceae

TYPE: Perennial, meadowlands, wetlands.

HABITAT: Marshes, meadows and brooks in sunny, wet locations.

DESCRIPTION: Groups form thick stands of tall, 2 m, plants with flat-topped flower heads on stiff purple stalks. Leaves are oblong attached in whorls. A major honey plant along wetland brooks.

FLOWER: Large flower head consisting of clusters of small, purple blooms with thread-like petals at top of main stalk. Florescence occurs July to September.

ZONE: 5a to 6b

HONEY INDEX: 3

Fagopyrum esculentum

Fallopia japonica

BUCKWHEAT *Fagopyrum esculentum*

TYPE:	Annual, meadowlands (field crop).
HABITAT:	On well-drained, acidic or low-nitrogen soils. Flourishes in cool weather and full sun.
DESCRIPTION:	Bushy, herbaceous plant 75 to 125 cm in height with rapid growth. Single, simple, arrow-shaped leaves. A major honey plant in temperate areas.
FLOWER:	Five-petalled pinkish-white flowers in clusters at end of stem. Bracts are green and pollen is ochre-red in colour. Bloom period occurs June to September.
ZONE:	4b to 6b
HONEY INDEX:	5 Distinctive, dark brown honey rich in antioxidants.

JAPANESE KNOTWEED *Fallopia japonica* *Polygonaceae*

TYPE:	Perennial, scrublands.
HABITAT:	Deep, wet wide ranging soils in ditches and fields. Prefers open areas in full sun. Fallopia overgrows waste areas quickly.
DESCRIPTION:	Rapidly growing knotweed reaches 3 to 4 meters height in one season. Extremely deep rhizome roots support broad oval leaves 15 to 20 cm long and 6 to 10 cm wide. Stems are hollow and segmented with nodes similar to bamboo. A major honey plant where abundant.
FLOWER:	Tiny white to cream coloured flowers appear on short stalks 15 cm long. Florescence occurs in late summer (Sept.) and persists for three weeks.
ZONE:	5a to 7a
HONEY INDEX:	3 Sometimes presented as 'Bamboo Honey'.

Fragaria virginiana

Gladiolus officinalis

STRAWBERRY

Fragaria virginiana
Rosaceae

TYPE:	Perennial, meadowlands.
HABITAT:	Prefers open fields on rich soil, often with grasses.
DESCRIPTION:	Short, spreading plant, 10 to 15 cm in height advancing by multiple runners spreading across the ground. A minor honey plant useful in providing springtime pollen for bees.
FLOWER:	Several flower stalks raise single white blossoms after the leaves emerge. Florescence occurs for three weeks in June.
ZONE:	5a to 7a
HONEY INDEX:	1

GLADIOLUS

Gladiolus officinalis

TYPE:	Annual bulb, landscape and garden flower.
HABITAT:	Rich loose soil in full sun.
DESCRIPTION:	Sixty to 120 cm in height; long lanceolate, upright leaves 4 to 5 cm in width emerge from root crown. Central florescence stalk rises above leaves. Gladiolus attracts nectar eaters from bees to hummingbirds. It propagates through seed and regeneration of bulbs.
FLOWER:	Tall central stalk supports large, brightly coloured, single bi-sexual flowers. Bloom appears sequentially from nearest the leaves to the top of the stalk. Three smooth or curled petals per flower 5 to 6 cm across. Flowering period is from July to September.
ZONE:	5b to 7a
HONEY INDEX:	2

Gleditsia spp.

Helianthus annuus

LOCUST (Black or Honey) *Gleditsia spp.*

TYPE: Perennial, woodland.

HABITAT: Large landscaped properties, parks and golf courses on rich, moist soils in full sun.

DESCRIPTION: Tall, spreading tree up to 35 meters in height; light green, alternate, compound leaves 20 to 30 cm long. Leaflets are oval, 4 to 5 cm long, shiny and dark on upper surface. Bark is dark brown, almost black and deeply furrowed on thick trunks. Seeds are encased in long, flat, dark brown pods. Usually a minor honey plant in Canada.

FLOWER: Occur in early July, after the leaves in long clusters of white, hooded blossoms 2 to 3 cm across. Bisexual and unisexual flowers occur on same tree. Duration of bloom two to three weeks.

ZONE: 5a to 7b

HONEY INDEX: 4
A premium honey.

SUNFLOWER *Helianthus annuus*

TYPE: Annual, garden and field crop.

HABITAT: Rich, well-drained soils in open sun.

DESCRIPTION: A single, tall, stiff stalk, 1 to 3 meters in height with large, alternate leaves 20 to 30 cm across. A single flower emerges from the apex of the hairy stalk. A major honey plant in cultivated areas.

FLOWER: A single, giant flower head appears in August. A single row of large, intensely yellow petals, 10 cm long, surround the florescence of hundreds of tiny, orange florets.

ZONE: 5a to 6b

HONEY INDEX: 4

Hesperis matronalis

Hieracium aurantiacum

SWEET ROCKET *Hesperis matronalis*

TYPE:	Perennial, meadowlands.
HABITAT:	Fields, ditches, roadsides, and gardens on mixed soils with good drainage in full sun.
DESCRIPTION:	Alternate, pointed and toothed leaves without petioles form a rosette just above the ground. A tall central stem with fine hairs reaches 50 cm. Stalked florescences emerge at each leaf junction along main stem to top. Generally a minor honey plant yet produces good quality nectar, attractive to bees.
FLOWER:	Lightly fragrant, four petalled in clusters on spikes along central stalk. Colour varies from plant to plant, in white, blue, purple and mauve; 2 to 3 cm in dimension. Main bloom six to eight weeks in June / July; extended late bloom until Sept.
ZONE:	5a to 7a
HONEY INDEX:	3

HAWKWEED *Hieracium aurantiacum*

TYPE:	Annual, meadowlands.
HABITAT:	Prefers dry, grassy areas in open sun.
DESCRIPTION:	Rosette of narrow leaves crown the root from which a hairy, thin, leafless stalk rises 30 to 60 cm, topped by a cluster of single flower heads.
FLOWER:	Multi-petalled, brightly orange and yellow single flower heads similar to sunflower, only 2 to 3 cm across. Primary bloom occurs in June and July.
ZONE:	4a to 5b
HONEY INDEX:	2

Hypericum perforatum

Kolkwitzia amabilis

ST. JOHN'S WORT — *Hypericum perforatum*, *Clusiaceae*

TYPE:	Annual, meadowlands and woodlands.
HABITAT:	Rough, gravelly soils in open sunny areas around fields, meadows and woods.
DESCRIPTION:	A central stalk, up to one meter, has many upright branches with small paired leaves and abundant flowers. May grow alone, or in clusters among other meadow plants. A minor honey plant for quantity, however considered a herbal or medicinal plant.
FLOWER:	Small, star-shaped flowers with five yellow petals. Pollen covered stamens protrude visibly from center of the blossom. An extended florescence period begins in July and lasts until September.
ZONE:	5a to 7a
HONEY INDEX:	2 (M)

BEAUTY BUSH — *Kolkwitzia amabilis*

TYPE:	Perennial, woodland / meadowland.
HABITAT:	Open, full sun sites in yards, fields, roadsides and landscapes on well-drained, mixed soils. Naturalized transplant for landscapes.
DESCRIPTION:	Large woody shrub to five meters in height, spreading 10 to 12 meters. Light green, simple, alternate oval leaves. A major honey plant where abundant. Highly attractive to Bumblebees as well.
FLOWER:	Large, pink bell-shaped flowers, 3 to 4 cm long, appear in June at the ends of long, weeping branches. Bloom period is two to three weeks.
ZONE:	5a to 7a
HONEY INDEX:	3

Lavendula spica

Lysimachia nummularia

LAVENDER

Lavendula spica
Lamiaceae

TYPE:	Meadowlands, perennial.
HABITAT:	Open, sunny areas on well-drained neutral soils. Herb gardens, rock gardens, flower beds and landscapes. It may occasionally be grown as a field crop.
DESCRIPTION:	Clusters of plants form thickets of thin stemmed, small leaved ground cover. A fragrant and aromatic member of the mint family. A major honey plant when cultivated.
FLOWER:	Numerous, small 'lavender' blue flowers cover the ends of the long, thin stems. Florescence occurs from July to Sept in multiple blooms.
ZONE:	5a to 7a
HONEY INDEX:	4 (M)

MONEYWORT

Lysimachia nummularia

TYPE:	Meadowlands and woodland, perennial.
HABITAT:	Coarse and rocky ground with good drainage in gardens, lawns and open areas. Prefers full sun.
DESCRIPTION:	Invasive, creeping vine with small round leaves that are smooth and shiny. A minor honey plant from limited growth and nectar.
FLOWERS:	Bright yellow, flat and shiny petals form 3 to 4 cm flowers along trailing vine. Florescence occurs in late summer for two to three weeks.
ZONE:	5a to 7a
HONEY INDEX:	1

Lythrum salicaria

Malus domestica

PURPLE LOOSESTRIFE

Lythrum salicaria
Lythraceae

TYPE: Meadowlands and wetlands, perennial.

HABITAT: Varied, from well-drained fields and roadsides to wet ditches and marshlands. Not particular about soil, but prefers open sunny areas.

DESCRIPTION: Plants vary in size from 30 cm to 85 cm in height; single or multiple stemmed. Leaves are lanceolate, 3 to 10 cm long and are in whorls of three. A major honey plant, but considered an invasive weed.

FLOWER: Four-petalled, purple flowers crowd upper half of vertical stalks, without stems.

ZONE: 5a to 7a

HONEY INDEX: 4
Produces premium honey.

APPLE

Malus domestica
Rosaceae

TYPE: Woodland (orchard), perennial.

HABITAT: Requires open sunny areas on well-drained, heavy soils.

DESCRIPTION: Medium to large, globular tree forming thickets. Leaves are oval and sharply toothed. Bark is rough and flakey, grey-brown with lichens. A minor or major honey plant depending on location.

FLOWER: Blooms before leaves appear in May - June with a prolific number of blossom clusters on short, thick twigs. The large 4-petalled flowers are pink and white with a strong fragrance and extremely attractive to honey bees. Florescence period of two to three weeks.

ZONE: 4b to 6a

HONEY INDEX: 3

Medicago sativa

Melilotus alba

ALFALFA

Medicago sativa
Fabaceae

TYPE:	Meadowlands, perennial.
HABITAT:	Open fields and pastures on deep soil.
DESCRIPTION:	Bushy herbaceous plant with large three-lobed leaves, up to 50 cm in height. A major honey plant producing a distinctive, premium honey.
FLOWER:	Pale to dark purple, single petal flowers clustered along thin, sinewy branches. Florescences from late June until September.
ZONE:	5a to 7a
HONEY INDEX:	5

WHITE SWEET CLOVER

Melilotus alba
Fabaceae

TYPE:	Meadowlands, biennial.
HABITAT:	Open sunny locations on any type of soil. Easily roots and grows in gravel and sand.
DESCRIPTION:	Tall, narrow and thin, up to 2.5 meters in height. Has small, narrow compound leaves on stiff, strong stems. Exhibits rapid growth in second year. A major honey plant producing lightly coloured, distinctive honey.
FLOWER:	Rows of tiny, pure white flowers emerge from thin upright branches and apex stem. Primary florescence occurs in June and July, continuing secondary bloom until September.
ZONE:	4b to 7a
HONEY INDEX:	4

Origanum vulgare

Oxalis stricta

OREGANO

Origanum vulgare
Lamiaceae

TYPE: Meadowlands, perennial.

HABITAT: Prefers open, sunny areas on well-drained soils that are neutral pH.

DESCRIPTION: Clusters of plants form thickets up to one meter in height. Leaves are alternate, rounded and light green. Stems are stiff and dark green to purple. Plant has fragrant oils that are released when rubbed. Generally a minor honey plant, however extremely potent as a medicinal herb for bees and people.

FLOWER: Numerous clusters of small, white to purple flowers cover the ends of the many upright stems. Florescence begins in mid-summer and continues strongly until the frost.

ZONE: 5b to 7a

HONEY INDEX: 2 (M)

YELLOW WOODSORREL

Oxalis stricta
Oxalidaceae

TYPE: Perennial, meadowlands.

HABITAT: On moist alkaline, sandy or loamy soils which are well drained. Prefers full sun and tolerates poor soils.

DESCRIPTION: Low lying, sprawling herbaceous plant with alternate leaves divided into three heart-shaped leaflet Mature seed capsules burst to expel seeds up to 4 meters. Attractive to bees.

FLOWER: Five-petalled, deep yellow flowers approximately 1 cm across, appear singularly on stems along the branches. Green pistols and bracts. Flowers are edible. Bloom period extends from early July to late October.

ZONE: 5a to 7a

HONEY INDEX: 1

Papaver spp.

Potentilla recta

GARDEN POPPY

Papaver spp.
Papaveraceae

TYPE: Annual, meadowlands.

HABITAT: Gardens, flower beds and open fields on rich, loose soils in full sun.

DESCRIPTION: Thin stemmed plants are 30 to 60 cm in height; leaves are heavily toothed and irregular 10 to 15 cm long, alternate without a petiole. Fruit consists of a hard capsule of seeds with a milky sap. A minor honey plant.

FLOWER: Beautifully bright and iridescent petals form single row or layers around corolla (species) of single flower atop each stalk. Florescence occurs in mid-summer lasting for several weeks.

ZONE: 5a to 7a

HONEY INDEX: 1 *

* Caution: some species are poisonous.

ROUGH-FRUITED CINQUEFOIL

Potentilla recta
Rosaceae

TYPE: Meadowlands, perennial.

HABITAT: Widely distributed in ditches, fields, and other marginal soils; prefers full sun on well-drained soils.

DESCRIPTION: Several stalks grow from a woody tap root to 80 cm in height; leaves are palmate, hairy and toothed on long stems, with a rough and hairy seed cap.

FLOWER: Five heart-shaped petals, pale white to cream are 2 to 2.5 cm across. Each stalk bears several flowers on individual stems. Florescence occurs in mid-summer and persists into the fall.

ZONE: 5a - 7a

HONEY INDEX: 1

Prunus armeniaca

Prunus avium

APRICOT

Prunus armeniaca
Prunus

TYPE:	Woodland, orchards, perennial.
HABITAT:	Wind protected sites in full sun on rich organic sandy soil.
DESCRIPTION:	Small to medium size tree with spreading branches and small oval leaves, dark green and smooth. Bark is reddish.
FLOWER:	Delicate, frost sensitive blossoms are first to appear in early spring before the leaves. Bracts are red and petals are pinkish white.
ZONE;	6a to 7a
HONEY INDEX:	1

SWEET CHERRY

Prunus avium
Prunus

TYPE:	Woodland, orchards, perennial.
HABITAT:	Open, sunny areas on well-drained fertile soils primary environment.
DESCRIPTION:	Medium to large globular tree with reddish bark on twigs. Leaves are oval and toothed.
FLOWER:	Clusters of white, five-petalled blossoms appear before the leaves in early spring; up to 2 cm across and very fragrant.
ZONE:	5b to 7a
HONEY INDEX:	2

Prunus domestica

Prunus persica

PLUM

Prunus domestica
Prunus

TYPE:	Woodland, orchards, perennial.
HABITAT:	Open areas in full sun, on moist limestone soils.
DESCRIPTION:	A medium size tree, spreading wide with many smaller branches. Leaves are alternate, simple and oval, sharp-pointed and course toothed. Bark is smooth grey-brown turning into thick, curling scales.
FLOWER:	May to June before or with the leaves, bisexual, five-petalled blossoms with green bracts. Profuse flowering occurs in small clusters of three or four blossoms. Flowering period of approximately two weeks.
ZONE:	5b - 6b
HONEY INDEX:	1

PEACH

Prunus persica
Prunus

TYPE:	Woodland, orchards, perennial.
HABITAT:	Thrives in open, full sun on fertile well-drained soils.
DESCRIPTION:	Small to medium size tree up to 4 meters in height. Distinctive, lanceolate leaves are mostly smooth edged. A minor honey plant with thin nectar.
FLOWER:	Numerous, fragrant, five petalled, pink blossoms appear before the leaves in early spring. The 2 cm flowers are borne on young twigs. Florescence occurs in April - May for two weeks.
ZONE:	6a to 7a
HONEY INDEX:	1

Prunus virginiana

Pyrus communis

CHOKE CHERRY *Prunus virginiana* *Prunus*

TYPE:	woodland, scrubland, perennial.
HABITAT:	Open forest areas, old fields and along roads and fences. Prefers rich, moist soils in full sun.
DESCRIPTION:	Small tree or shrub with reddish-brown bark. Leaves are finely toothed and oval, light green, 5 to 7 cm long. A major honey plant.
FLOWER:	Profuse clusters on thick stems of white five petalled blossoms 1 to 1.5 cm across. Bisexual, they bloom May / June for three weeks.
ZONE:	5b to 7a
HONEY INDEX:	4
	Aromatic, flavourful honey.

PEAR *Pyrus communis*

TYPE:	Woodland, orchards, perennial.
HABITAT:	Moist, rich soil with good drainage in full sun.
DESCRIPTION:	Sturdy, upright tree is conical-shaped. Leaves are oval, light green and smooth. Bark is brown and smooth
FLOWER:	Clusters of bright, white blossoms cover the tree before leaves in May / June. Bisexual blossoms open for approximately two weeks.
ZONE:	5a to 7a
HONEY INDEX:	1

Rhododendron spp.

Rosa rugosa 'Hansa'

RHODODENDRON *Rhododendron spp.*

TYPE:	Perennial, woodland.
HABITAT:	Tolerant of shade, prefers full sun on acid soil. Found in flower beds, lawns and gardens.
DESCRIPTION:	Small to medium woody, broadleaf shrub (both deciduous and evergreen varieties) used as an ornamental landscape bush.
FLOWER:	Very large, abundant blooms that persist for several weeks.
ZONE:	5a to 7a
HONEY INDEX:	0 Although there is no scientific evidence to support the premise that this plant produces toxic honey, it is rumoured as such and is otherwise an insignificant source of nectar. Avoid if possible.

HANSA ROSE *Rosa rugosa 'Hansa'*

TYPE:	Perennial, pasture lands.
HABITAT:	Flower beds, gardens, lawns and some naturalization in abandoned areas. Rosa prefers full sun on rich organic well-drained soils that are slightly acidic.
DESCRIPTION:	A hardy shrub rose of an older variety. A large spreading bush that reaches 2 meters by 2 meters. Leaves are compound, medium green and toothed. Bark is brown and thorny. A minor honey plant although producing very high quality.
FLOWER:	Conspicuous, large pink blooms appear in early July. Typical rose shape with circular petals. Flowering period six to eight weeks.
ZONE:	5a to 7a
HONEY INDEX:	4

Rosa rugosa 'Henry Hudson'

Rosa multiflora

HENRY HUDSON ROSE *Rosa rugosa 'Henry Hudson'*

TYPE: Perennial, pastureland.

HABITAT: Old farms, naturalized along roads and fences on mixed soils. Prefer full sun and well-drained, slightly acid soil.

DESCRIPTION: Bushy shrub, less than a meter in height and across. Medium green compound leaves are toothed. Bark is dark, almost black and thorny. A minor honey plant that produces good pollen for bees.

FLOWER: Large, white flowers with a single row of petals, 4 cm across.

ZONE: 5a to 6b

HONEY INDEX: 3

TEA ROSE *Rosa multiflora* *Rosaceae*

TYPE: Perennial, woodland, scrubland.

HABITAT: In full sun on soils rich in organic matter, slightly acidic and well-drained. Glower beds, gardens and landscapes with some wild naturalization.

DESCRIPTION: An intensely cultivated species with a large number of flora varieties. Leaves are compound oblong and sharply toothed. Depending on variety heights are from 50 cm to more than three meters. All varieties are woody and thorny.

FLOWER: Very large, multi-petalled blooms in a wide range of colours and textures. Very fragrant with typically yellow pollen. Bloom period may be short or long (up to three months). A major honey plant when abundant that produces a dark, aromatic honey.

ZONE: 4b to 7a

HONEY INDEX: 5

Rosa virginiana

Rosa rugosa spp.

WILD ROSE *Rosa virginiana*

TYPE:	Perennials, meadowlands.
HABITAT:	Prefers sunny, open locations on heavy soils.
DESCRIPTION:	A multi-stemmed woody plant, thorny, 1.3m in height forming thickets. Leaves are small, oval and shiny. Rosa is a major honey plant producing flavourful darker honey.
FLOWER:	Large pink to red petals, 5cm across, bright yellow center. Bloom occurs throughout July.
ZONE:	4b to 7a
HONEY INDEX:	4

HARDY SHRUB ROSE *Rosa rugosa spp.*

TYPE:	Perennial, pastureland.
HABITAT:	Flower beds, lawns, gardens and naturalized along fences and roads. Tolerates mixed soils and partial shade.
DESCRIPTION:	Hardy, spreading shrub grows to three meters in all directions with long branches drooping in a floral cascade. Leaves are light green small and toothed. Bark is brown and thorny. A minor honey plant unless abundant.
FLOWER:	Mostly pure white, simple flowers 2 to 3cm across with a single row of petals. Has profuse florescence across the ends of nearly all branches. Bloom period four to five weeks.
ZONE:	4b to 6b
HONEY INDEX:	4 All species of rose produce very aromatic and flavourful honey. It is worth seeking out large areas dominated by the Rosa rugosa.

Rubus fruticosus

Rubus spp.

BLACKBERRY *Rubus fruticosus*

TYPE: Perennial, woodland.

HABITAT: In woodcuts, abandoned fields and fence rows on heavy soil.

DESCRIPTION: Spreading, upright, thorny canes up to 2 meters in height with compound leaves of 3 to 5 leaflets. A major honey plant which produce a high quality honey, light coloured honey..

FLOWER: Clusters of large, showy, white blossoms on long stalks appear on stiff, two year old canes. Florescence occurs in June / July for about three weeks.

ZONE: 4b to 7a

HONEY INDEX: 4

DEWBERRY *Rubus spp.*

TYPE: Perennial, woody, meadowlands.

HABITAT: In abandoned fields, along fences and roadsides on mixed soils rich in woody material. Prefers full sun, will tolerate partials shade.

DESCRIPTION: Sprawling, woody canes trail across the ground to a height of 30 to 40 cm. Compound leaves of five oblong course leaflets. Berries are larger and sweeter than blackberries; a major honey plant when in abundance.

FLOWER: Large, five-petalled brilliant white blossoms appear June / July (just after cane blackberries begin to bloom) in clusters on new twigs up to 12 cm long. Stamens show dark pollen. Florescence lasts approximately two to three weeks.

ZONE: 5a to 7a

HONEY INDEX: 4
Rubus produces an excellent, premium, light coloured honey.

Solidago spp.

Sorbus decora

GOLDENROD *Solidago spp.*

TYPE:	Perennial, meadowlands.
HABITAT:	Prefers sunny locations on new ground, fields and roadsides.
DESCRIPTION:	A stiff central stalk, 1m height, supports thin branches on the upper portion of the plant. Leaves are lanceolate and long. This major honey plant produces large amounts of quality honey.
FLOWER:	Tiny, yellow flowers crowd along thin branches. Florescence occurs in August and September.
ZONE:	4a to 7a
HONEY INDEX:	5

SHOWY MOUNTAIN ASH *Sorbus decora*

TYPE:	Perennial, woodland.
HABITAT:	Cool mountain woods, edges of streams and lakes, on moist soil.
DESCRIPTION:	A small, bushy tree 5 to 12 meters in height with an oval shape. Sorbus grows singly or in clumps in mixed forests. The bark is greyish-brown to grey and mostly smooth. It produces clumps of hard red berries in late summer that persist late into the season.
FLOWER:	Appear in June, after the leaves in broad, flat-topped clusters of small, white bisexual blossoms. Bloom period is about two weeks.
ZONE:	4a to 7a
HONEY INDEX:	1 Rich source of pollen for wild, native bees and honey bees.

Stellaia graminea

Syringa vulgaris

STITCHWORT

Stellaia graminea
Caryophyllaceae

TYPE:	Annual, meadowlands.
HABITAT:	Pastures, gardens, lawns and fields.
DESCRIPTION:	A small, discrete plant has lanceolate leaves easily blending in with its surroundings. A minor honey plant due to size and distribution.
FLOWER:	Small, star-like five petalled flowers, 1cm across, deeply divided to the base develop on thin stems. Florescence occurs throughout summer.
ZONE:	5a to 6b
HONEY INDEX:	1

LILAC

Syringa vulgaris

TYPE:	Perennial, woodland.
HABITAT:	Prefers open, sunny areas on rich soil.
DESCRIPTION:	Small to medium bushy tree with deep, dense roots and wood. A major honey plant that produces fragrant honey.
FLOWER:	Large clusters of small flowers on individual stalks on outer branches. Colour varies from white to purple. Bloom lasts three weeks in June.
ZONE:	5b to 7a
HONEY INDEX:	3

Taraxacum officinale

Thymus vulgaris

DANDELION

Taraxacum officinale
Asteraceae

TYPE: Annual, meadowlands.

HABITAT: Prefers open, sunny areas on any soil.

DESCRIPTION: Ubiquitous, fast growing spreading plant with long, jagged leaves. Major honey plant, produces harvestable honey of quality and is an important food source for bees.

FLOWER: Single yellow flower on each of multiple stalks emanating from tap root. Florescence May to June up to six weeks.

ZONE: 4a to 7a

HONEY INDEX: 4

THYME

Thymus vulgaris

TYPE: Perennial, meadowlands.

HABITAT: Prefers open, sunny areas on well-drained soils. Herb gardens, flower beds and borders of landscape areas.

DESCRIPTION: Short ground cover, 10 to 15cm in height, thickly branched with thin stems and tiny round, green leaves and very aromatic when brushed. A minor honey plant with special attributes.

FLOWER: Tiny, light-blue flowers, 2 to 4mm across are abundant along thin stems. Florescence begins in early July and continues for six weeks.

ZONE: 5a to 7a

HONEY INDEX: 2 (M)
Rich with aromatic oils and antibiotic phenols, the nectar is beneficial to bee health and human nutrition.

Trifolium repens

Trifolium hybridum

WHITE CLOVER *Trifolium repens*

TYPE: Annual, meadowlands.

HABITAT: Exposed ground, short grasslands, pastures.

DESCRIPTION: A short spreading ground cover, 10 to 15cm in height, with three leaved stems. A major honey plant produces most common honey.

FLOWER: Tall stems are topped with white florets in a globular shape. Main florescence occurs in June and July followed by intermittent flowering to September.

ZONE: 4b to 7a

HONEY INDEX: 4

ALSIKE CLOVER *Trifolium hybridum*

TYPE: Annual, meadowlands.

HABITAT: In open fields and pastures on rich to mixed soils. Alsike thrives among grasses and other wildflowers.

DESCRIPTION: A low growing, up to 25cm in height, spreading meadow/pasture plant named for its rounded three lobed compound leaves. A major honey plant produces distinctive honey.

FLOWER: A rounded array of florets atop a single stalk. Flowers are medium sized, 1 to 2cm a across, pink blush on white in colour. Bloom begins in June and runs for six weeks or more.

ZONE: 4b to 7a

HONEY INDEX: 5
A hybrid cross between white ground clover and red clover, produces abundant quantities of high grade, light coloured honey.

Trifolium spp.

Tussilago farfara

RED CLOVER *Trifolium spp.*

TYPE:	Annual, meadowlands.
HABITAT:	In open fields and pastures on rich to mixed soils. Trifolium thrives among grasses and other wildflowers.
DESCRIPTION:	A low growing, up to 40cm in height, spreading plant characterized by three lobed compound leaves. A minor honey plant for honey bees.
FLOWER:	Single florescence atop individual stalk consisting of a rounded array of florets. Flowers are rouge to deep red; very large 2-3cm in diameter with an extended bloom period. Blossoms will reappear after mowing.
ZONE:	4b to 7a
HONEY INDEX:	2 Due to the depth of the florets, the nectar, which is abundant, is not available to the honey bee's short tongue. Cutting reduces the size of the flowers.

COLTSFOOT *Tussilago farfara*

TYPE:	Annual, meadowlands.
HABITAT:	In open, sunny areas in rocky, sandy ground.
DESCRIPTION:	Flower stalks appear before small, oblong leaves, approximately 15cm height. Minor honey plant, however is extremely valuable for first nectar in spring. Also is an important source of fresh pollen for growing bee colonies.
FLOWER:	Single yellow flower on single stalk. Bloom April to May.
ZONE:	4b to 6b
HONEY INDEX:	1

Vaccinium angustifolium

Vaccinium macrocarpon

BLUEBERRY *Vaccinium angustifolium*

TYPE: Perennial, meadowlands.

HABITAT: Prefers peaty, high acid soils on well drained slopes in full sun.

DESCRIPTION: Short woody, spreading ground cover up to 30cm leaves are smooth, alternate and oval; light green leaves have coppery tinge. A minor honey plant for honey bees.

FLOWER: Small, whitish, bell shaped flowers distributed along spreading branches. Bloom period occurs in June lasting for two weeks. Some varieties blossom later than others

ZONE: 4a to 6b

HONEY INDEX: 2
While honey bees will visit the blossoms, they are not well suited to pollinate this plant and collect meagre quantities of honey.

CRANBERRY *Vaccinium macrocarpon* *Ericaceae*

TYPE: Perennial, wetlands.

HABITAT: Tidal bogs that flood seasonally; grows in sandy soils in open sun.

DESCRIPTION: Low, creeping shrubs up to 2 meters wide and 20 cm in height. Leaves are small, evergreen 10 to 20 mm long. Fruit is large, red, slightly acid. A minor honey plant, though plentiful.

FLOWER: Pink with four petals reflexed outward to fully expose the pistol and stamens. Small blossoms, 5 to 7 mm, proliferate along the upper vines. Florescence from June to July for 3 to 4 weeks.

ZONE: 4a to 6b

HONEY INDEX: 1

Verbascum thapsus

Vicia cracca

COMMON MULLIEN *Verbascum thapsus*

TYPE:	Annual, meadowlands.
HABITAT:	In gardens, on open ground and ditches.
DESCRIPTION:	A single stalked plant, 1.5m height, with large drooping leaves below the inflorescence. A minor honey plant despite great size.
FLOWER:	Yellow, five petalled flowers emerge on all sides of the upper stalk and branches. Florescence occurs form July to September.
ZONE:	5a to 7a
HONEY INDEX:	1

TUFTED VETCH *Vicia cracca*

TYPE:	Annual, meadowlands.
HABITAT:	Fields and pastures on fertile soil.
DESCRIPTION:	Rapidly growing vine in hot weather with small paired leaves on separate branches. A major honey plant produces abundant flavourful honey.
FLOWER:	Clusters of purplish, tubular flowers hang from a climbing vine. Florescence occurs June, July and August.
ZONE:	5a to 7a
HONEY INDEX:	5

PLANTING BEE PASTURE

Although the common literature claims it is inefficient to grow bee pasture, I think that much can be achieved through planting bee pasture.

As well as developing specific honey types by melliferous plant selection, I like to use honey plants in landscape design. With one broad stroke you are able to generate select honey, pollinate fruits and vegetables, provide habitat and nutrition to indigenous pollinating insects, and build beautiful ornamental gardens and landscapes.

With careful planning and plant selection from the honey plant guide, you can create flowering, insect paradises in almost any location and of various sizes.

An important consideration in regards to any bee pasture is the duration of the nectar flow. To properly support the bees throughout the year, and prepare them for surplus honey production sufficient nectar must be available throughout their life cycle. In most rural settings of Eastern Canada this factor is provided for by the indigenous flora growing in the fields, forests and river banks. Having verified a sustainable nectar and pollen supply, one can proceed to design and establish the secondary or supplementary bee pasture. These can vary in size, type and bloom period depending on your particular goals and location.

Before designing can begin an examination of the local conditions is necessary. Parameters necessary to successful design are climatic variables, soil type(s), exposure to wind and sun, and the lifestyle of the owner. Practical and aesthetic considerations should have approximately equal weight in decisions.

There will be controllable elements such as the economics of material cost and labour, the choice of plants and also the fixed elements such as lot size, shape and grade. Once a budget is estimated and considered acceptable the joyful work of design can begin.

The design can be very simple, as in the seeding of a lawn with white clover (Trifolium), or intensely complicated as in a flower bed that blooms with different flowers continuously throughout the season.

Composition is the most important principle of landscape design. This is most commonly achieved through the use of a theme. While the main implied theme of planting bee pasture is the feeding of honey bees, bumblebees, and other pollinating insects (butterflies), a secondary theme based on human aesthetics can be adopted.

A good landscape relies on balance, rhythm and focus. The theme is created or represented by the balance of repeated elements. Let's use the rose-garden as our theme. In this case the central focus would be on the Rosa genus utilizing the many species of rose to highlight the theme. Rhythm would be created by the spacing of the plants in either regular symmetrical fashion for a 'formal' display, or an irregular asymmetrical fashion for the 'informal' garden.

A well designed landscape will have a focal point, which can be an inanimate object such as a large stone, or a fountain, or it can be a particular plant of interest such as a tree. Patterns in the planting should emanate from or to the focal point. Once the theme and focal point have been established, the other more subtle elements can be introduced. These considerations would be the colour scheme, texture and background of the layout. Using the rose theme again, there are multiple colours of flowers which should be considered as if painting a canvas for effect. Roses also come in a number of shapes and sizes, such as bush, creeping and climbing. The garden then is designed in layers to achieve the desired effect of balance, rhythm and focus. Other plants may be interspersed throughout the roses to provide contrast and background.

And, in keeping with our main objective to provide bee pasture, it should be noted that roses are an excellent honey plant both for the bees and the select honey they make.

Using the Quick Calendar chart, plant species can be chosen

to bloom in sequence throughout the season providing the bees with a rich and consistent food supply, and offering the grower a splendid display of colour, scent and activity all summer long. Mixed wildflower gardens and fields are easy to prepare and offer a great abundance of variety, colour, and bee activity.

Although most of the flowering trees bloom for a relatively short time, they produce excellent grade honey and present striking landscapes, whether as focal points, border areas or groves. Consider the size, shape and form of plants when they are fully grown. For instance, a grove of locust trees surrounded by tea roses and wildflowers will grow into a balanced layered bee pasture of extreme beauty.

It is a common mistake for many beginners or amateurs to plant sapling trees too close to a building. Soon a huge maple tree is pushing its roots into the basement and its branches through the bedroom window.

Trees and bushes have particular shapes and densities. These shapes range from conical, columnar, round, and vase-like to prostrate. Referring back to balance again, these shapes should not be haphazardly mixed. A good landscape should be layered with tall specimens in the background, medium sized and shaped plants next and short, wide foreground flora. The density refers to the sun blocking surface area of the leaves. 'Shade trees' have dense leaf coverage and allow very little sunlight to reach the ground beneath them. This limits the number and type of plants that will grow under them. The majority of honey plants prefer full sun and would not prosper under shade. Under light density trees, such as locust, other honey plants can be positioned next or under them.

Thus we have the principles of fine landscape design. Keep in mind the use of unity of type, simplicity in execution, variety in colour and texture and finally the pleasing effect of balance.

Examples:

Flower beds can display scintillating colours in a sequential bloom that begins in April and continues throughout the summer into the frosts of fall. Begin with the early blooming bulbs of Crocus, Daffodil and Tulip. Plant in spaced groups for best effect. An excellent choice for May flowers that attract numerous species of bees is the Heath (Erica carnea). For June and July the choices become wider and beautiful nectar plants such as Sweet Rocket and Poppies provide excellent adornment. As the season turns toward late summer and autumn the Asters are both attractive and productive.

It should be noted that certain common 'household' flowers should not be considered for bee pasture. Plants that produce natural insect inhibitors and repellents do not promote a bee friendly environment. These would include, although not limited to, the Chrysanthemum, Geranium, Marigolds, and Nicotiana.

Fruit and vegetable gardens offer an excellent opportunity to grow a sumptuous array of delicious food for the family and nurture the bee population. If gardens are of sufficient size, a crop of select honey is also possible. The legume family is very productive both in fruit and nectar. Consider a variety of beans, peas and vetches. Peppers and tomatoes will also benefit from an abundance of bees. Although squashes, melons and pumpkins require pollination they are mostly favoured by bumblebees.

Roadsides and ditches can be made attractive and productive (provided excess road salt is not present) with seeding and transplanting of 'invasive' and spreading honey plants. Raspberries, Blackberries and Dewberries work well. Vetches and clovers are also easily established.

Acres and fields can be mechanically seeded with Alfalfa, wildflowers and herbs such as dill, oregano, and mint. Various types of orchards may be planted in suitable areas. Cherries, plums, apples, and pears produce excellent honey and fruit. Allowing the undergrowth of wildflowers in an orchard enhances both the natural appeal and environmental conditions for bee health and prosperity.

With imagination, an open mind and the specifics of nectar and honey plants beautiful, productive landscapes and positive environments are possible.

Seeding Transitional and Marginal Lands:

We often encounter abandoned or unused lands nearby or in the neighbourhood that can easily be transformed into brilliant bee pasture. Assuming there is no objections from landowners (seeking permission is recommended), we can purchase kilograms of useful seed from the local 'Farmer's Co-op' to re-plant the unproductive land.

Unless the property is contaminated with petrochemicals or nuclear radiation, or is composed of solid bedrock, it can be re-habilitated. The first step is to assess the condition and permeability of the soil. A quick flip of the ground with a spade will reveal the depth, condition and moisture content of the soil. Using this information the correct starter plant for the rehabilitation can be chosen. Even if presented with an apparently barren, stoney terrain the land can support nectar plants.

Abandoned stone quarries are an excellent example and site to replenish with productive, flowering plants. Using the honey plant guide, the soil and habitat can be matched to the best nectar plants to initiate the re-greening of the otherwise waste land.

As the plants establish themselves, and soil develops other nectar plants of more mature tolerances can be added to grow out over time a rich, lush and highly productive bee pasture.

The following examples illustrate how to proceed:

In the case of gravel pits, railroad lines, and other loose stone sites the primary or pioneer plant to use is the Sweet Clover (Melilotus alba, Melilotus officinalis). Their deep tap roots establish themselves well, produce copious quantities of nectar and the seeds are widely available.

In the case of clay and sand soils also sweet clover should be used, but can also be augmented with Vetch (Vicia sativa) and Alfalfa (Medicago sativa). These seeds are also widely available.

In the case of better soils, mixed wildflowers, Scotch and Canada Thistles, Fireweed and Sweet Rocket can be introduced. The seeds for these plants would most likely have to be collected from existing bee pasture the previous year.

It is a good practice to harvest seed from the plants of productive bee pasture for use in other places and to develop new sites.

Birds are a valuable asset to the beekeeper, and anyone who wishes to promote bee pasture. In particular the small seed and fruit eating birds, such as finches, and warblers distribute the seeds of many plants across a wide area. The serviceberries, black and raspberries, and blueberries are well distributed and propagated by the droppings of these songbirds. In growing and promoting bee pasture a wide range of flora and fauna benefit to the joy of all.

BLOOM CALENDAR

	Apr	*May*	*Jun*	*Jul*	*Aug*	*Sep*	*Oct*
Alfalfa			X	X	X		
Alsike Clover			X	X	X		
Apple			X				
Apricot	X	X					
Aster					X	X	X
Beautybush			X				
Blackberry				X			
Blueberry			X				
Borage				X	X	X	X
Buckwheat			X	X	X	X	
Burdock				X	X		
Canada Thistle			X	X	X		
Chicory				X	X	X	X
Chives				X			
Choke Cherry		X					
Coltsfoot	X						
Common Mullien				X	X		
Cranberry				X	X		
Crocus	X						
Dandelion		X	X				

	Apr	May	Jun	Jul	Aug	Sep	Oct
Dewberry				X			
Dogwood				X			
Fireweed				X	X	X	
Garden Poppy			X	X			
Gladiolus				X	X	X	
Goldenrod					X	X	
Hansa Rose				X	X		
Hardy Shrub Rose				X	X		
Hawkweed			X	X			
Hawthorn			X				
Henry Hudson Rose				X	X		
Heath		X					
Horse Chestnut			X				
Japanese Knotweed						X	
Joe-pye Weed				X	X		
Lavender				X	X	X	
Lilac		X					
Locust (Black or Honey)				X			
Moneywort					X		
Northern Catalpa				X			
Oregano				X	X	X	X
Peach		X					

	Apr	May	Jun	Jul	Aug	Sep	Oct
Pear			X				
Plum		X					
Poppy				X	X		
Pumpkin				X	X		
Purple Loosestrife				X	X		
Queen Ann's Lace				X	X		
Red Clover			X	X	X	X	
Red Maple	X						
Rhododendron				X	X		
Rough-Fruited Cinquefoil				X	X	X	
Rhododendron				X	X		
St. John's Wort				X	X	X	
Serviceberry		X					
Scotch Heather					X		
Scotch Thistle				X	X		
Showy Mountain Ash				X			
Speckled Alder	X						
Stitchwort			X	X			
Strawberry			X				
Sugar Maple		X					
Sunflower					X		
Sweet Cherry		X					

	Apr	*May*	*Jun*	*Jul*	*Aug*	*Sep*	*Oct*
Sweet Clover				X	X	X	
Sweet Rocket			X	X			
Tea Rose				X	X		
Thyme				X	X		
Tuffted Vetch			X	X	X		
Wild Rose				X			
White Clover			X	X	X		
White Sweet Clover			X	X	X	X	
Yellow Rocket		X					
Yellow Woodsorrel				X	X	X	X

TOXIC PLANTS

Poisonous plants exist in abundance around the world. From the skin irritating oils of poison oak and ivy, to the oxcylic acid bearing leaves of the common rhubarb to the deadly fruits of the nightshade and the hallucinogenic peyote cactus extends a mighty family of plantae that defend themselves with toxic chemicals. While other plants use means of sharp thorns, stinging hairs, and waxy or thick bark to protect their lives against browsing animals.

Most angiosperms have evolved to counter predation by production of excessive quantities of seed. In the primitive flora, this is seen in the copious quantities of pollen that are produced and released into the air. In the higher plants, involving cross-pollination and insect pollination they have adapted by producing vast quantities of pollen and insect attractants such as sugar rich nectar, intense scents and an elaborate array of colours and patterns.

At first blush, it would seem completely self-defeating for a plant species to produce nectar that would harm or kill the insect pollinator. In the complex biological network, 'the struggle to survive' has produced numerous examples of seemingly contradictory modes of reproduction and defence. Thus, caution must be exercised in regards to potentially toxic melliferous plants.

A very important element in the assessment and appraisal of any bee pasture is the proximity and quantity of potential poisonous plants. These plants may be indigenous or transplanted species for ornamental purposes. Geographical location will have the largest influence on the type and quantity of nefarious flora to any given bee pasture. Some areas of the world would be completely unsuitable for honey production, while others are exceptionally good. However, good areas may be spoiled by the introduction of transplanted gardens of dangerous, toxic plant species. Therefore, careful knowledgeable surveys of potential bee pastures should be made before and during honey production, and beekeeping operations are conducted.

Should toxic plants be discovered in any area within the bees' foraging area (up to five kilometres from their hive) decisions must be made to mitigate the risk. The honey producer would have three possible choices when encountering dangerous flora. In the case of large tracts of toxic plants the operator would practice avoidance (either by geographic location or by timing of bloom periods). The alternate method would be the removal of all toxic plants (or their flowers). The second method is only feasible if there are small numbers of plants and all their locations are known. Lastly, no attempts are made to alter the landscape; such as when a neighbour has one Azalea in their garden. (Observations of bee activity are recommended.)

In regards to serious human poisoning by nectar and honey in Eastern Canada the risk is extremely low as this area has small numbers of indigenous toxic melliferous plants.

Caution is warranted for honey producers near peat bogs that may contain the Bog-rosemary.

There is one other kind of 'poisonous' plant nectar. That is of the offensive scent. A prime example is the Hawthorne shrub, Crataegus spp., which has small, bright white flowers in the early spring and attracts many bees. Upon inspection the blossoms emit an unpleasant odour that is repulsive to humans, but acceptable to hungry honey bees. If followed through to the bee hive one can extract honey with the very same foul odour. However, it does not appear to be debilitating to bees or animals.

The last form of toxic plant is the one engineered by humans. These plants come in two forms. First is the common crop which is systematically sprayed with petrochemical herbicides and pesticides -- such as the deadly neonicitinoids. This chemical causes deterioration of the nervous system and results in a slow death for the bees and the colony. They often have enough time to transport some quantities of this poison to the honey stores within the hive before they succumb.

And the other toxic plant is the GMO (genetically modified organism) which is a transgenic hybrid plant (combining of genes from different species) which produces a new toxin imbedded within its DNA.

Of these types of unnatural plants all are dangerous (especially to the honey bee) and none is necessary.

POISONOUS MELLIFEROUS PLANTS

Common Name	Linnaeus	Geographic	Toxicity
Azalea Rhododendron	Rhododendron ponticum	U.K., S. & E. Europe	Honey & Plant
Bog-Rosemary	Andromeda polifolia	Asia, Europe, North America	Honey & Plant (lethal)
Calico Bush Mountain Laurel	Kalmia latifolia Kalmia angustifolia	Northern & Eastern U.S.A.	Honey & Plant
California Buckeye	Aesculus californica	Western U.S.A.	Toxic to Honey bees
Datura 'Thorn Apple'	Datura stramonium	Mexico; East Europe	Honey & Plant (lethal)
Deadly Nightshade Belladonna	Atropa belladonna	North America; Europe	Honey & Plant (lethal)
Deathcamases Star Lily	Zigadenus glaberrimus	South East U.S.A.	Toxic to Honey bees
Evening trumpet flower; Woodbine	Gelsemium sempervirens	Temperate to tropical America	Toxic to Honey bees
Henbane	Hyoseymus niger	Eastern Europe	Honey & Plant (lethal)
Oleander	Nerium oleander	Southern Europe & North Africa	Honey & Plant (lethal)
Opium Poppy	Papaver somniferum	World-wide temperate zones	Honey & Sap (narcotic)
'Soapberry'	Serjonia lethalis	Brazil	Honey & Plant (lethal)
Tutu	Coriaria arborea	New Zealand	Honey & Plant (lethal)
Wharangi Bush	Melicope ternata	New Zealand	Honey & Plant

PLANT HARDINESS ZONES

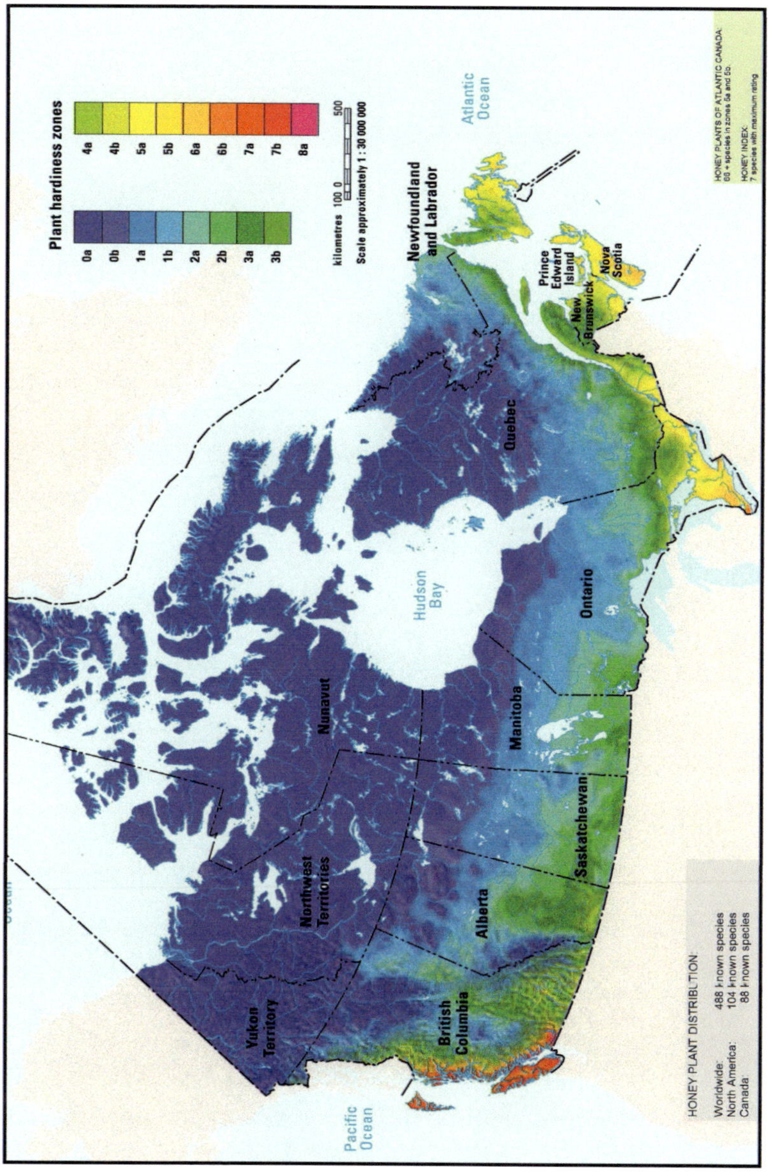

POST SCRIPT

The preceding list of Honey Plants is not botanically exhaustive. While the Major honey plants have been described, many other marginal and minor honey plants of interest are clearly absent.

Presently work continues at data collection to enhance and expand the knowledge base for Canadian flora of significance to bee culture.

Future editions of Honey Plants of (Eastern) Canada intends to provide complete contextual documentation of all known melliferous botanicals.

Made in the USA
Monee, IL
05 November 2026

67385914R00067